Gold Placer Deposits of the Pioneer District Montana

By J. T. PARDEE

CONTRIBUTIONS TO ECONOMIC GEOLOGY, 1951

GEOLOGICAL SURVEY BULLETIN 978-C

A geologic study of the historic diggings along Gold Creek

UNITED STATES DEPARTMENT OF THE INTERIOR

Oscar L. Chapman, *Secretary*

GEOLOGICAL SURVEY

W. E. Wrather, *Director*

For sale by the Superintendent of Documents, U. S. Government Printing Office
Washington 25, D. C. - Price 45 cents (paper cover)

CONTENTS

ILLUSTRATIONS

GOLD PLACER DEPOSITS OF THE PIONEER DISTRICT, MONTANA

By J. T. Pardee

ABSTRACT

The first discovery of gold in Montana was made in 1852 in the alluvium along Gold Creek in the Pioneer district. The deposit was low grade, but in time richer gravels were found nearby. Little mining was done before 1870, owing to the scarcity of water and the lure of higher grade diggings elsewhere. In 1870, however, abundant water was made available through a ditch from Rock Creek, and in the next two decades the richer deposits were largely worked out. Since 1890, mining activity has varied in intensity, but has ceased entirely for only short periods. Since 1934, low-lying deposits have been partly mined by dredging. At the time of the last field work in connection with this examination (1949), only small-scale sluicing was in progress, dredging was suspended, and only one company (the Master Mining Co.) was operating with more than moderate success.

The Pioneer district is an indefinitely bounded area of about 80 square miles which includes the northern end of the Flint Creek Range and the adjoining benchlands of Deer Lodge Valley. The mountains are chiefly of Mesozoic sedimentary rocks, an intruded mass of Tertiary granite, and, marginal to the granite, a zone of metamorphic rocks. In places these rocks are covered by Pleistocene glacial deposits and later alluvium. The benchlands are underlain by Cretaceous sedimentary rocks, Tertiary volcanic rocks and "lake beds," and Quaternary glacial and alluvial deposits.

In late Tertiary or early Quaternary time streams eroded a plain across the benchlands that sloped gradually from the mountains toward the interior of Deer Lodge Valley. In early Pleistocene time, glaciers, descending from the northern part of the Flint Creek Range, deposited gold-bearing drift on the adjoining part of the plain. In the following interglacial time the streams cut valleys into the plain, deepened them, step by step, and left gravel-covered terraces marking each halt in the down cutting. During this process most of the drift and its gold was concentrated into gravels that underlie the terraces.

In two later stages, glaciers again descended to the benchlands and deposited the intermediate drift containing moderate amounts of gold, and the late drift containing relatively little gold.

In post-Pleistocene time, parts of the older deposits were reworked by streams and the gold reconcentrated in alluvium along their channels. Although mining has exhausted the gold reserves of much of the terrace gravel and later alluvium, some known gold placer reserves are contained in alluvium along lower Pioneer Gulch, in patches of terrace gravel, and remnants of the early glacial drift. Probable gold reserves include alluvium along lower Gold Creek, patches of terrace gravel, and parts of the intermediate glacial drift. Some possible gold reserves may exist in parts of the deposits of late glacial outwash.

INTRODUCTION

This examination of the Pioneer (Gold Creek) district (fig. 26) was made by the U. S. Geological Survey to determine the occurrence, character, and extent of the gold placer deposits, including those deposits currently too low grade to be of commercial value, but which, under changed conditions and improved mining methods, may prove to be of economic value.

FIELD WORK

The field work began in 1916, when J. T. Pardee, assisted by T. H. Rosenkranz, started making a sketch map of the benchlands (pl. 11) using a plane table and open-sight alidade. During that season the central and southern parts of the area were mapped. Mapping of the remainder of the area was continued intermittently through 1939, by method of compass traverses. Brief field checking was done in 1949 by R. H. Thurston. During the course of the mapping, mines which were operated at different times were seen in various stages of development.

ACKNOWLEDGMENTS

Valuable information was given by Conrad Kohrs, who had played a leading part in the early development of the district and by Frank J. Slaughtner, who, for many years, conducted the hydraulic operations at the Kohrs and Bielenberg mine. Credit is due T. H. Rosenkranz for careful assistance in the topographic mapping of the benchlands, and to Newton Cleaveland, superintendent of the Pioneer Dredging Co. from 1934 to 1939, and other mine operators for their cordial cooperation and assistance. To R. H. Thurston, who visited the district in 1949, the writer is indebted for criticism of the manuscript and for corrections and additions to the maps.

PREVIOUS INVESTIGATIONS

Before the present survey began, little detailed information was available concerning the extent and geologic relations of the deposits within the benchlands. However, most of that part of the district in the Flint Creek Range (fig. 26) is within the Philipsburg quadrangle, previously mapped and described by Emmons and Calkins (1913, pp. 76, 79, 81, 87, 108, 263, and pls. 1, 2).

LOCATION AND ACCESSIBILITY

The Pioneer district (fig. 26) is an area of about 80 square miles of the benchlands in the southwestern side of Deer Lodge Valley. It is within Powell and Granite Counties, western Montana and is about 14 miles northwest of Deer Lodge, the county seat of Powell County. The town of Goldcreek, on the Northern Pacific Railway and the Mil-

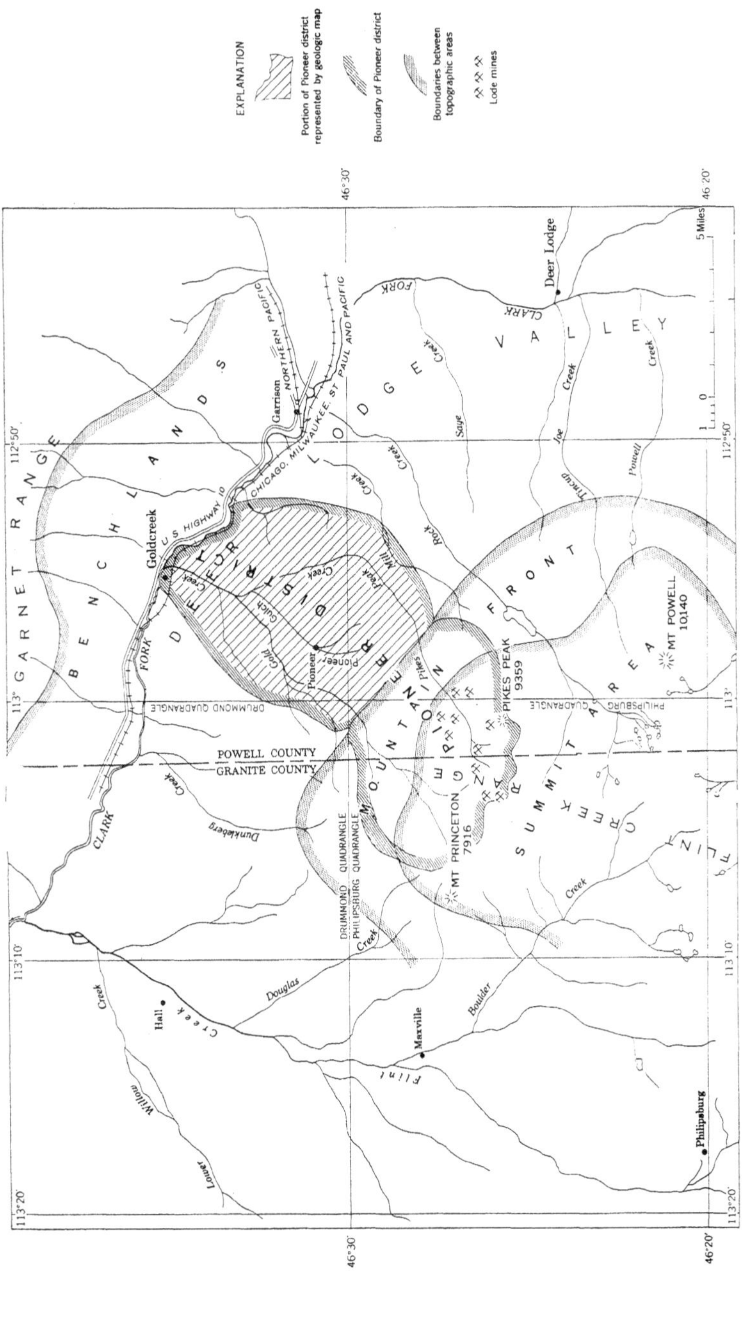

FIGURE 26.—Map of Pioneer district and adjacent areas.

waukee (Chicago, Milwaukee, St. Paul & Pacific) Railroad, is a short distance north of the district and less than a mile south of U. S. Highway No. 10. The part of the town of Pioneer that has escaped destruction by dredge mining is 5 miles by road south of Goldcreek. From Deer Lodge, Pioneer may be reached also by a secondary road that, for part of the way, follows the route of a former military highway known as the Mullan Road.

TOPOGRAPHY

The larger part, about 45 square miles, of the Pioneer district (fig. 26) is in the benchlands of the southwestern side of the Deer Lodge Valley; the remainder extends into the northern part of the Flint Creek Range.

"Benchlands" is the local name for areas characterized by spurs or benches that project toward the interior of the basin and end in escarpments overlooking a lowland along the main stream. (See pl. 11.) These benches form such prominent features as Windy Hill, a broad spur that has a tabular summit. It is east of Pikes Peak Creek and about 4½ miles from the mountain front. Another spur west of Pikes Peak Creek is broad at the north and greatly narrowed southward. Gold Hill (altitude 5,600 feet) and two smaller summits, Hill 5400 East and Hill 5400 West, are on this spur. Ballard Hill rises to a summit of 6,000 feet on the short spur between Pioneer Gulch and Gold Creek. The spur west of Gold Creek that is transected by Griffin Creek is comparatively low except for a prominent summit, Benetsee Butte (altitude 5,000 feet), at its northern end. The spurs are separated by the valleys of streams that descend from the mountains toward the middle of the basins where they join the main streams. The floor of the Gold Creek valley is about half a mile to a mile in width. It is formed of low flats on flood plains and shows the uneven surfaces characteristic of glacial moraines and glacial outwash. The sides of these valleys, and also the ends of the spurs, are generally terraced.

In contrast to the benchlands, the area within the Flint Creek Range consists chiefly of steep-sided rocky canyons, glacial cirques, and rugged peaks. In places the main divides are formed by wide, flat or gently sloping surfaces at the general summit altitude of 8,000 to 8,500 feet. Parts of the cirque floors 1,000 to 2,000 feet below are also flat or moderately uneven. This mountainous part of the Pioneer district (fig. 26) is mostly within the Philipsburg quadrangle.

DRAINAGE

The Pioneer district is a part of Deer Lodge Valley, a broad intermontane basin enclosed by the Flint Creek Range on the south and west, the Garnet Range on the north, and the Continental Divide on the east. The trunk stream, the Clark Fork of the Columbia River, locally known as the Deer Lodge River, follows a northward course near the middle of the basin and escapes toward the northwest through a gorge (beyond the area covered by the map, fig. 26).

Most of the Pioneer district is drained by Gold Creek and its principal tributary, Pikes Peak Creek. Gold Creek enters the Clark Fork about a mile southeast of the town of Goldcreek. Pikes Peak Creek joins Gold Creek about 2 miles above the mouth of the latter. A central part of the benchlands is drained by Pioneer Gulch which enters Pikes Peak Creek a short distance above its junction with Gold Creek. Relatively small northeastern and southeastern parts of the benchlands are drained, respectively by Independence Creek and Mill Creek; these streams enter the Clark Fork above (southeast of) Gold Creek. In its course of 8 miles between the mountain front and the Clark Fork, Gold Creek descends about 1,000 feet. The headwaters of Gold Creek and Pikes Peak Creek drain the mountainous part of the district.

VEGETATION

The lower slopes of the Flint Creek Range are, in most places, thickly covered by a coniferous forest, and much of the summit area is above timberline. The benchlands is a grass-covered area suitable for grazing and, where irrigation is available, for cultivating crops. It lacks timber except for the cottonwoods and various species of poplar that grow along the streams.

HISTORY OF MINING

Inasmuch as the discovery and the first mining of gold in Montana took place within what is now the Pioneer district, the early history of that district is of more than ordinary interest.

The first discovery of gold in Montana was made in 1852 by an Indian half-breed, Francois Finlay. The discovery is mentioned briefly by Browne (1867, pp. 496–497) and by Stuart (1876, p. 6) who also describes the part he played in the subsequent exploration and development of the deposits.

Finlay's prospecting did not reveal deposits rich enough to mine. News of his discovery, however, gradually circulated among the few traders or others in the region and in 1856 the party of Robert Hereford prospected in the district. In 1858 Stuart and others found gold and, again in 1860 and 1861 parties led by Stuart found promising placer ground in some of the dry gulches tributary to a stream first called Benetsee Creek, later Gold Creek.

In the summer of 1860 Henry Thomas, better known as Gold Tom, found a little gold all the way down in a 30-foot shaft sunk in the glacial debris along Gold Creek at a point about a mile west of the site that became Pioneer City, the early gold mining camp. The following summer he recovered gold amounting to about $1.50 a day from some of the surface material which he washed in small axe-hewn sluices.

Stuart's party planned to mine the deposits they had discovered, but, because of a lack of supplies, operations were delayed until May 8, 1862, when they set out their sluices and began to mine. These sluices, excepting the small operation of Gold Tom, were the first ever used in Montana.

During 1862, many gold seekers reached the Pioneer district. One party arriving about May 14 found diggings on Pioneer Gulch yielding gold worth 20 cents to the pan (Langford, 1902). Another party arriving about May 20 prospected successfully along a branch of Gold Creek which they named Pikes Peak Creek. Still another party was observed by Mullan (1863, p. 34), who passed that way about June 1, to be "recovering about 10 dollars a day to the man" by the method of sluicing.

The locality of Finlay's discovery was probably at or near the crossing of Gold Creek and an old Indian trail, the trail that later was followed by the U. S. Military highway known as the Mullan Road. This point is a short distance above the mouth of the stream about a mile southwest of the present town of Goldcreek (pl. 11). A settlement of brief duration, called American Fork, which was established at the crossing about 1860, has been identified by the remains of cabins built by Stuart and others.[1]

During the summer of 1860, many gold-seekers were headed for the new field. Some stopped to prospect on the way, and in August a party from Salt Lake City, led by John White, found the placer gravels at Bannack, Montana, which were "so rich that other mines were abandoned, and by fall nearly all the miners in the Territory had congregated there" (Browne, 1868, p. 498).

Col. G. W. Morse,[2] a prominent citizen of Granite County, reported that in August 1862 he saw only two miners at work in the Pioneer district. They were mining the gravel along Gold Creek just below Pikes Peak Creek, and were recovering gold worth $1.50 to $2.00 per day to the man by the method of "shovelling in," using sluice boxes made of whip-sawed lumber.

The next six years was a period of successive rich discoveries, and the rapid mining of gold throughout western Montana. Raids by organized robbers were common, and the Vigilantes were finally organized to control the prevalent lawlessness. The Vigilantes brought the first law and order to the Territory with the capture and

---

[1] From a talk by W. A. Clark at a meeting of the Montana Pioneers, Oct. 1, 1929, published in the Montana Standard, a Butte, Mont., daily newspaper; issue of Oct. 7, 1929.

[2] Personal communication.

execution of the bandits and robbers who were preying upon the gold shipments from producing camps. While discoveries in other districts were being made in quick succession, Pioneer remained almost deserted; when, figuratively speaking, "the cream had been skimmed" from richer deposits, the miners began to return to Pioneer.

From the start, mining in the Pioneer district had been hindered by the lack of sufficient readily available water for the elevated terraces or the heavier bouldery deposits below. This difficulty was overcome by the construction of the Rock Creek ditch (a canal some 16 miles long) in 1868–69 by Conrad Kohrs and others. The ditch delivered water at a point on Gold Hill above nearly all the terraces. The first of these terraces to be extensively mined were on the slope descending to Pikes Peak Creek. Soon several hundred miners were at work. The terraces contained rich deposits and gold worth as much as $140,000 was recovered in a single season from part of the pit excavated on Batterton Bar (pl. 11). A mining camp, called Yam Hill, thus came into existence.

By 1874 or 1875, the richer parts of the terraces had been worked out and most of the miners had gone to the vicinity of Pioneer where rich "diggings" had been found in a series of terraces on a spur called Pioneer Bar and on Ballard Hill. When additional water was brought in from Gold Creek, mining of these deposits continued for several years, during which Yam Hill became practically deserted and Pioneer "City" a more substantial settlement.

More than a million dollars worth of gold (based upon $20.67 a fine ounce) is said to have been washed in the late 1870's and early eighties from the terrace gravels on Pioneer Bar (pl. 11).

A part of the early drift on the west side of Gold Hill was mined before 1920 in a working called the Squaw Gulch pit. In 1920 the Rock Creek ditch and its water rights were purchased by agricultural interests and the water was diverted to use outside the Pioneer district. Water brought from Pikes Peak Creek was later used to mine deposits on Gold Hill above the Rock Creek ditch. The largest of these workings, known as the Kelley and Irvine pits, were made in early glacial drift at the south of the Squaw Gulch pit. For several years after 1920, mining was on a small scale with the exception of a fairly extensive hydraulic operation carried on by Slaughtner in the gravel of Pioneer Gulch along the foot of Ballard Hill. This deposit had been by-passed during previous operations because of a heavy talus overburden.

A group of Chinese miners recovered an undetermined but possibly large amount of gold before 1930 from deposits along the upper parts of the middle and east forks and Pioneer Gulch, from the terrace east of lower Gold Creek known as China Bar (fig. 27), and from

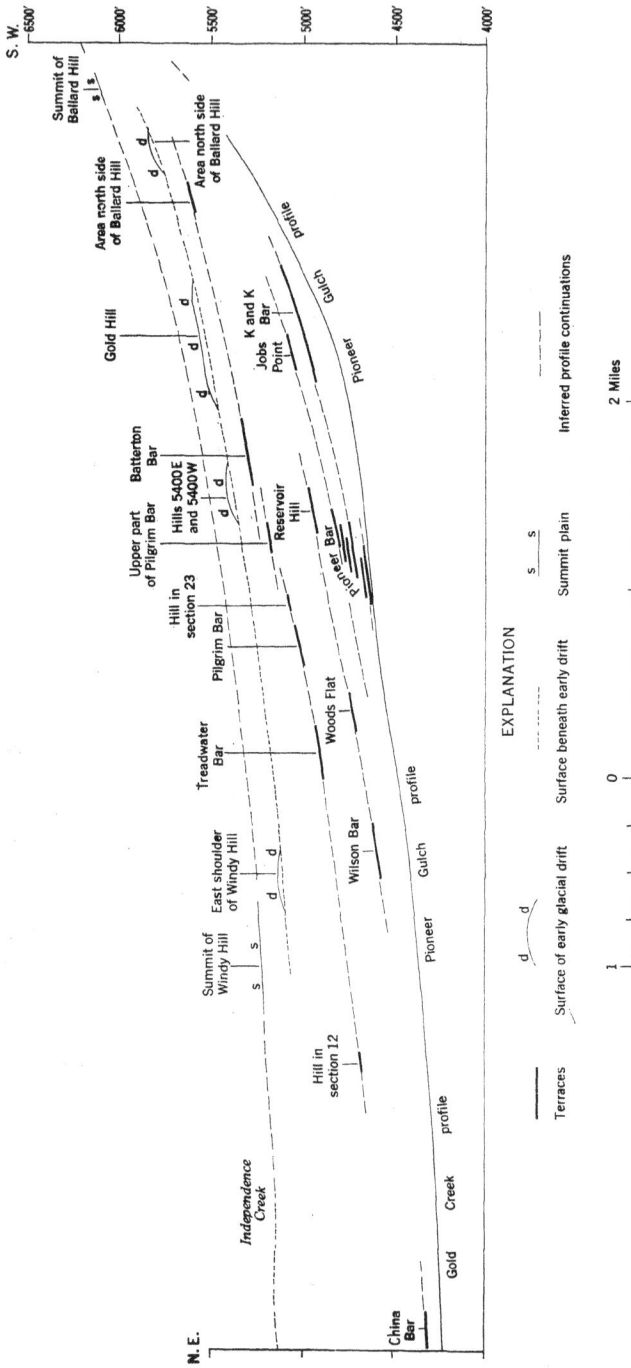

FIGURE 27.—Projected profiles of the terraces.

abandoned mines elsewhere. In 1930–31 the Henderson Mining Co. prospected the gravel in the lower part of Squaw Gulch (pl. 11), but the results were said to be unsatisfactory.

About the beginning of the present century an English company unsuccessfully attempted dredge mining in the valley of Gold Creek at a point known as the old dredge pit just below the mouth of Pioneer Gulch (pl. 11). The dredge was too small and unsubstantial for the deep, bouldery gravel, and was abandoned after having excavated a shallow trench. The site of this operation (old dredge pit, pl. 11) was in a large tract then owned or controlled by Carl Bergstrand. Later the tract was acquired by Pat Wall.

Before 1890 rather large pits were mined on the southeast shoulder of Windy Hill east of Pikes Peak Creek (see pl. 11) with water supplied by a branch of the Rock Creek ditch. The operations are said to have been profitable but the amount of gold recovered is not known.

By 1890 Pioneer Bar and the deposit at the Ballard mine had been largely worked out, but extensive hydraulic operations (fig. 28) were underway on K & K Bar, a low terrace in Pioneer Gulch. Except for seasonal interruptions, mining of this deposit was continued until 1918 by F. J. Slaughtner, for the owners, Kohrs and Bielenberg.

By the summer of 1916 the workings had reached a point about a mile and a half above Pioneer. The width of the mined area ranges from 200 feet to about 1,000 feet, and is greatest where it expands into two large pits near the upper end. (See pl. 11.) The deposit ranged in thickness from a few feet near the edge of the terrace to 50 feet toward the middle. As the mining progressed upstream, the excava-

FIGURE 28.—Hydraulic operation at Kohrs and Bielenberg mine on K & K Bar.

tion downstream became filled with tailings. Operations in 1916 extended the working face upstream from 40 to 90 feet along a width of about 600 feet.

In December 1933, the Pioneer Placer Dredging Co., under the direction of Newton Cleaveland, after making preliminary tests of the deposits and obtaining possession or control of an area embracing most of Pioneer Gulch and the Bergstrand-Wall property in the adjoining part of Gold Creek valley, began mining with a sturdy California-type electrically operated chain-bucket dredge (fig. 29). By

FIGURE 29.—Gold-mining dredge of the Pioneer Placer Dredging Co. in operation along Pioneer Gulch.

the time operations were suspended in 1941, a pit from 100 to 400 yards wide and about 3,000 yards long (pl. 11) had been mined to an average depth of about 8 yards. The total area of the pit was about 930,000 square yards; its volume was nearly 7,500,000 cubic yards, and the reported recovery of gold was about 18 cents a yard from material mined.

A gravel deposit at the head of the middle fork of Pioneer Gulch, actually the continuation of a gold-bearing streak earlier mined by the Chinese, was explored during the summer of 1935 by Conrad Warren and Andrew Moore. At the same time, J. J. Ramer and Joseph Boucher prospected a deposit along the west fork of Pioneer Gulch. Gold placers were mined at times in the valley of Mill Creek, though mostly on a small scale. F. C. Howard and others operated a small hydraulic plant at the Orphan Boy mine during 1935 (pl. 11).

In the mountainous part of the Pioneer district (fig. 26) rather extensive placer (hydraulic, bulldozer, and power shovel) operations have been carried on along the banks of upper Gold Creek over a distance of about 2 miles. Beginning at a point about 5 miles by road upstream from Pioneer (approximately 2 miles beyond the southwest limit of the area shown on pl. 11) and continuing southwestward, the principal placer workings are known as the Wisner, Tibbets, McFarland, Pineau, and Hughes placers. One or more of these placers has been worked nearly every season since 1890 except during World War II. In August 1949 the Master Mining Co. was operating mines in the Wisner, Tibbets, and McFarland properties.

The Pineau placer, also known as the Friday mine, was one of the largest of the upper placers, and was operated for many years by Gus Pineau. The Silver Cloud Mining Co. operated the Pineau placer during part of its period of production. The production of the upper placers is not known with certainty, but the Pineau workings yielded the greater part of the gold mined on upper Gold Creek before 1941. The production of the Pineau before 1926 is not known, although it was presumably large, but from 1926 to 1931 the Pineau mine was the chief source of the $19,343 production credited to the Pioneer district (U. S. Bureau of Mines). Its total production to 1935 has been popularly estimated at more than $50,000. Records are lacking for the other mines of the vicinity with the exception of the McFarland mine which is said to have produced $35,000 (Emmons and Calkins, 1913, pp. 263–264).

GEOLOGY

THE MOUNTAINOUS PART OF THE DISTRICT

Most of the mountainous part of the Pioneer district (fig. 26) is within the Philipsburg quadrangle and has been described by Emmons and Calkins (1913, pp. 70–88). The remainder extends into the adjoining area on the east, which has not been geologically mapped. The upper Gold Creek placers (Wisner, Tibbets, McFarland, Pineau, and Hughes) are within the Philipsburg quadrangle.

In the mountainous area pre-Tertiary sedimentary rocks, ranging from the Madison limestone of Carboniferous age to the Colorado formation of Upper Cretaceous age, are folded into north-striking anticlines and synclines, and are intruded by Tertiary (?) basic sills, and Tertiary biotite granite. The Phosphoria formation (Permian) is mapped with the Quadrant quartzite (Emmons and Calkins, 1913, pl. 1). The Phosphoria is exposed at the head of Gold Creek and in a glacial cirque near the headwaters of the north fork of Pikes Peak Creek.

TERTIARY INTRUSIVE ROCKS

Sills of diorite and diorite porphyry (Emmons and Calkins, 1913, pp. 80–88, pl. 1) intruded into the Colorado formation, are exposed in the basin of Gold Creek a short distance south of the north line of the Philipsburg quadrangle. Along the south side of the basin a large body of biotite granite (Emmons and Calkins, 1913, pp. 108–111), the Royal batholith, has invaded about one-third of the mountainous part of the district.

As a rule, the sedimentary rocks are extensively metamorphosed where they occur close to the intrusive bodies, particularly the sandstones and shales which may be so highly metamorphosed as to form varicolored flint-like rocks called hornstones, and quartz-mica schists. A conspicuous variety of altered rock is andalusite schist derived from the black shale of the Colorado formation (Emmons and Calkins, 1913, pp. 76, 79, 81).

QUATERNARY GLACIAL DRIFT AND ALLUVIUM

Quaternary glacial drift occupies considerable areas in the valleys and cirques. It rests on bedrock eroded, mainly by glaciers, across all of the different formations mentioned. Associated with the drift are small patches of Recent alluvium (Emmons and Calkins, 1913, pl. 1).

The study of the rocks included in the various glacial drifts shows that the material was derived from the northern portion of the Flint Creek Range, and that in successive invasions, the glaciers cut deeper and deeper into this range. The early drift is characterized by a preponderance of the younger sedimentary rocks, particularly the quartzites, and also of metamorphosed sedimentary rocks, the hornstones and schists. The intermediate glacial drift perhaps contains a higher percentage of metamorphic rock fragments than either the early or late drift. The material deposited during the last glacial invasion is composed principally of fresh granite cobbles and boulders that originally occurred at depth in the headwater basins of Gold Creek and Pikes Peak Creek. The percentage composition of metamorphic rocks, sedimentary rocks, and granite in the glacial deposits may be used as a guide to determining the age of any particular deposit belonging to one of the three glacial invasions.

THE BENCHLANDS

PRE-TERTIARY ROCKS

Sandstones, shales, and conglomerates, all probably of Cretaceous age, underlie Ballard Hill, most of the spur west of Gold Creek valley, and small areas along Willow Creek.

TERTIARY VOLCANIC ROCKS

Tertiary andesitic lava and breccia form a broken chain of low hills along the lowland of the Clark Fork (pl. 11). The volcanic rocks rest on an eroded surface of older sedimentary rocks, and are overlain unconformably by Tertiary "lake beds."

TERTIARY "LAKE BEDS"

Tertiary "lake beds" form the bedrock in most of the placer mines (pl. 11), and are exposed in several of the steeper slopes and terrace escarpments. The "lake beds" are composed mostly of beds of mixed volcanic ash and clay, alternating with thinner beds of ash, marl, and sand or fine gravel. A wash-out below a ditch on Reservoir Hill exposed 100 feet of pale buff mixed ash and clay in distinct, nearly horizontal beds from 2 inches to a foot thick. Some of the beds are brownish from included organic material.

About 100 feet of nearly horizontal "lake beds" are exposed in section 14, on the southeast side of Perryman Hill. They are of pale buff mixed ash and clay interbedded with a few thin layers of ash. A carbonaceous layer occurs near the bottom of the exposure. A bed near the top contains small snail-like fossil shells. Wm. H. Dall,[3] of the U. S. Geological Survey, identified a *Lymnaea* and 25 species of *Planorbis* from this bed. His conclusion is that the age of the beds is probably middle Miocene.

The "lake beds" have been eroded from the valley of Gold Creek and from most of the spur along the west side of the valley (pl. 11). East of the Gold Creek valley, however, the lake beds, as indicated by exposures in mine pits and terrace escarpments, are continuous throughout most of the benchlands.

LATE TERTIARY OR EARLY QUATERNARY GRAVEL

Within the benchlands the oldest of the formations that overlie the Tertiary lake beds is a stream laid gravel that covers the summits of Windy Hill and Ballard Hill (pl. 11). These gravels have not been mapped because they are patchy and irregularly distributed. These patches of gravel are the remnants of a formerly extensive sheet that formed a plain descending from the mountains at a height of 700 to 800 feet above the present valley of Gold Creek (fig. 27). The gravel on Ballard Hill is 30 to 40 feet thick and resembles a coarse torrential wash. On Windy Hill, 4 miles from the mountain front, it is thinner, and consists of a medium- to fine-textured washed and sorted gravel. Its boulders are predominantly of quartzite, but it contains a few of quartz-mica schist and some of highly siliceous rock, called hornstone (see p. 80).

No striated boulders or other evidence of glacial origin was observed in this gravel. Its relative elevation (fig. 27) above any of the glacial

[3] Personal communication.

deposits indicates that it is the oldest gravel of the area. Benetsee Butte (pl. 11) is also capped with a gravel of similar composition and at the same relative elevation, but its correlation with the deposit on Windy Hill was not determined with certainty.

QUATERNARY FORMATIONS

The Quaternary formations of the benchlands comprise an early glacial drift, terrace gravel, intermediate glacial drift, late glacial drift, and alluvium.

Early glacial drift.—Remnants of an early glacial drift cover relatively small areas (pl. 11) on Gold Hill, on the southeast shoulder of Windy Hill (fig. 30), the summits of Hill 5400 East, Hill 5400 West, and the north slope of Ballard Hill.

FIGURE 30.—Early glacial drift southeast of Windy Hill. The exposed boulders and cobbles are of quartzite.

The early drift is not more than 20 feet thick where exposed in mine workings on Gold Hill; it is deeply weathered, and consists of an unsorted mass of boulders, cobbles, and clay. It includes irregular lenses of washed gravel. Ninety percent of the boulders and cobbles are quartzite; the remainder are mostly hornstone and a very few are fine-grained granite. The quartzite boulders are angular or subangular and some are as much as 4 feet in diameter (fig. 31). The early drift at the surface on Hill 5400 East, Hill 5400 West, and on the southeast of Windy Hill is similar in lithology to that on Gold Hill. A few boulders of diorite were observed on Hill 5400 West.

FIGURE 31.—Quartzite boulders from early glacial drift on Gold Hill.

The boulders in the drift on Ballard Hill are about half of quartzite, and half of andalusite schist, other metamorphic rocks, and granite. Glacial striae are preserved on many of the schist boulders.

Terrace gravel.—Stream-washed gravel caps the terraces and some summits correlated with the terraces, namely: Jobs Point, Reservoir Hill, and hills in the southern parts of sections 12 and 23 (pl. 11). The gravel contains many 6- and 8-inch cobbles and boulders from 1 to 3 feet across where it occurs near areas of early drift, but becomes finer textured away from such drift. The gravel is usually unconsolidated except for a thin layer near the surface which is generally cemented with calcium carbonate (caliche). The terrace gravel occurs in thicknesses ranging from 1 to 12 feet where it is exposed in many mine workings.

Intermediate glacial drifts.—The intermediate glacial drift covers areas of moderate extent along Mill Creek, at the head of Pioneer Gulch, and west of Gold Creek immediately below the mountain front. It occurs also in the K & K Bar (fig. 30) beneath the late glacial drift, and may occur in areas beneath the late drift along Gold Creek.

The intermediate drift contains abundant boulders of granite and quartzite; the granite is iron-stained and partly decayed. The deposit west of Gold Creek includes a few boulders of diorite.

Late glacial drift.—The late glacial drift underlies two relatively large areas (pl. 11). One deposit extends down the Gold Creek valley to about the center of section 21; the other occupies Pikes Peak Creek for 3 miles below the mountains and extends into the upper part of the basin formed by Squaw Gulch and the upper part of Pioneer Gulch. The late drift is composed largely of fresh granite boulders

and only a minor amount of quartzite or other rocks.   In both areas
the drift occurs in lateral and terminal moraines, and in the Gold
Creek valley and Squaw Gulch in outwash that lies below the moraines.
A prominent lateral moraine separates French Gulch and Gold Creek,
and a lateral moraine which swings down from the mountain front,
forms the divide between Pikes Peak Creek and the Squaw Gulch-
Pioneer Gulch basin.

Late glacial outwash in the Squaw Gulch-Pioneer Gulch basin extends about a mile below the lateral moraine and is about 10 feet thick
where exposed in the Kohrs and Bielenberg mine (fig. 32), near the

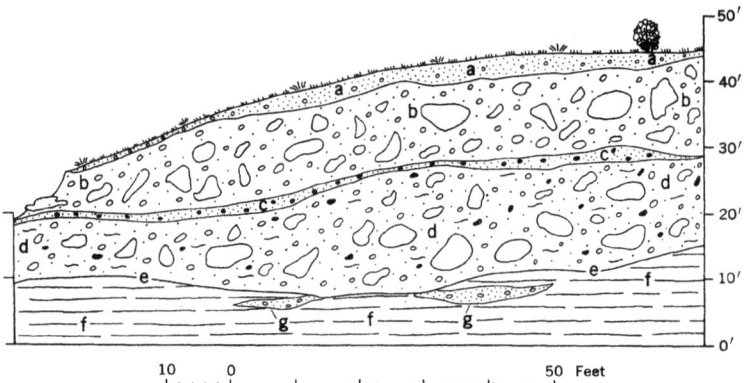

FIGURE 32.—Northeast face of the 1916 pit of the Kohrs and Bielenberg mine. a, sand
and pebbles ; b, late glacial outwash with fresh granite boulders ; c, fine-textured gold-
bearing gravel ; d, intermediate glacial outwash with rusty and decayed granite boulders ;
e, eroded bedrock surface ; f, Tertiary "lake beds" dipping 15° SW. (toward the ob-
server) ; g, fine gravel with angular pebbles.

upper end of the K & K Bar.   The sides of older mine pits, partly con-
cealed talus, show that the outwash is at least 50 feet thick at the back
of the terrace and near the middle of the ridge between upper Pioneer
Gulch and Squaw Gulch.   Late glacial outwash in the Gold Creek
valley forms a belt a quarter of a mile to half a mile wide extending
from the area of moraines about 3 miles downstream.

Alluvium.—Quaternary alluvium covers an area 100 to 1,000 yards
wide and about 6 miles long that extends downstream from Pioneer to
the mouth of Gold Creek where it merges with the alluvial deposit
along the Clark Fork (pl. 11).   The alluvium covers smaller areas in
Pioneer Gulch above Pioneer, in Pikes Peak Creek above its mouth,
in Squaw Gulch, along the forks of Mill Creek, and in several
smaller gulches.   Alluvium also forms a few scattered deposits, not
mapped, which overlie late glacial drift within areas occupied chiefly
by moraines.

The Recent alluvium is rather coarse and packed with boulders near
the mountains and close to glacial deposits.   As the distance from the

mountains and glacial deposits is increased, the boulders are less abundant and alluvium becomes finer textured. Away from the mountains the upper layers of alluvium are composed mostly of clay, sand, and fine gravel but commonly coarse gravels are present at depth.

PHYSIOGRAPHY

The development of the benchlands of the Pioneer district and their associated placer deposits, are closely related to the physiographic history of the general region.

Deer Lodge Valley is one of several intermontane basins that are widely distributed throughout western Montana. The basins and the surrounding mountains were formed chiefly by local differential depressions and uplifts of the surface that accompanied a general elevation of the region in late Tertiary and early Quaternary time (Pardee, 1950, p. 359, pl. 1).

During the Eocene epoch, erosion reduced the region (northern Rocky Mountain province) to a generally low surface of small to moderate relief. The free drainage swept away practically all the land waste.

In the following Oligocene epoch and part of the Miocene epoch, certain areas that were roughly equivalent to the present intermontane basins were slightly depressed by crustal movements. At times the depressed areas were ponded and at times thin blankets of volcanic ash were showered over the region. Ash that fell in the ponds was preserved unmixed. Ash that fell on surrounding lands was washed into the depressions with some debris from the land. The Tertiary "lake beds" are therefore composed mostly of clay, sand and gravel mixed with ash, some interbedded layers of ash and, occasionally, a bed of fossiliferous marl.

In a succeeding period of time that extended into the Pleistocene another general elevation of the region occurred. This movement was accompanied by uplifts of the present mountains and deformation in the basins that to a varying extent faulted and tilted the "lake beds." Early in this period the trunk streams that drain the basins were established in their present courses. As the mountains were uplifted these streams deepened their channels as fast as the mountains rose.

In late Tertiary or early Quaternary time, gravels were deposited on plains that were eroded across the "lake beds." The plains sloped from the mountains toward the interior of the basins where they merged along the trunk stream. Later on as the basin outlets were lowered, the streams cut valleys into the plains, deepened the valleys step by step, and left graveled terraces to mark each pause in the downcutting. The dissected areas, locally called benchlands, occupy broad belts along the sides of the basins. They are characterized by spurs (benches) that project from the mountains between the streams and end in escarpments overlooking the lowlands along the main stream. Remnants of the late Tertiary or early Quaternary plain occur on some spurs. The benchland area of the Pioneer district has been dissected to a depth of 700 to 800 feet below the late Tertiary or early Quaternary plain as represented by the summits of Ballard Hill and Windy Hill (fig. 27). Shallow valleys eroded in the plain by the ancestral Pikes Peak Creek and the ancestral Gold Creek, are represented by surfaces beneath the early drift on Gold Hill, on the east shoulder of Windy Hill, and in other elevated positions (fig. 27). Remnants of later valley floors form the successively lower terraces down to and including China Bar, which is east of Gold Creek near its mouth, and the series of terraces on the spur, known as Pioneer Bar, west of Pioneer Gulch.

The mountainous part of the Pioneer district is limited to the headwater basins of Gold Creek and Pikes Peak Creek in the northern part of the Flint Creek Range. The mountain front, rising steeply from the saddle behind Ballard Hill, is probably the worn scarp of a fault. The fault trends southeastward, traces the boundary between the benchlands and the mountain (fig. 26, and pl. 11), and has elevated the mountain block. Elsewhere the general features of the range, including its dome-like form (Pardee, 1950, p. 402) indicate that the range was elevated chiefly by warping of the late Tertiary surface.

The dissection of the mountains, concurrent with that of the benchlands, was effected in the mountains partly by glaciation. Three times glaciers scoured and deepened the headwater basins of Gold Creek and Pikes Peak Creek and descended onto the benchlands. At the time of the first glacial invasion, only shallow valleys had been cut into the late Tertiary or early Quaternary plain (fig. 27). At the time of the last invasion dissection of the benchlands had reached its present depth.

GOLD PLACER DEPOSITS

DEPOSITS IN THE BENCHLANDS

Placer deposits of known or probable value are widely distributed in the benchlands. They include parts of the Quaternary formations mapped (pl. 11) as early glacial drift, terrace gravels, intermediate drift, late glacial outwash, and alluvium. The placer gravels were concentrated by erosion from drift transported by glaciers. This drift was transported from an area of gold-bearing lodes in the Flint Creek Range to the adjacent benchlands in the Deer Lodge Valley.

GOLD HILL

The early drift on the western side of Gold Hill, above the steep slope to Squaw Gulch, has been mined in extensive workings (pl. 11), the northern or main part of which was called the Squaw Gulch pit, and the southern and smaller part, the Kelley and Irvine pits.

Where exposed in the south end of the Kelley and Irvine pits the drift is mostly an unsorted mass of boulders, cobbles, sand, and clay. Some of the boulders are 3 or 4 feet in average diameter. Much larger ones are said to have been uncovered in mining. The bouldery mass incloses irregular lenslike bodies of gravel from 5 to 12 feet thick that are stained brownish in the upper part and red in the lower part by iron oxides. The gold occurs chiefly as small, rather smooth grains in the iron-stained material, called by the miners the "blood-red gravel." Elsewhere the originally vertical sides of the workings are crumbled down to talus slopes that conceal details of structure, but not the general character and composition of the drift.

Streaks of gold-bearing gravel occur in small gulches and ravines on Gold Hill. These deposits have been concentrated from the early drift by intermittent streams originating from rainstorms and melting snows.

A total of about $20,000 worth of gold is reported to have been recovered from the Kelley and Irvine pits, and the Squaw Gulch pit is said to have yielded a much larger, although unknown amount.

HILLS 5400 EAST AND 5400 WEST

The patches of early drift on Hill 5400 East and Hill 5400 West are similar in surface exposures to the drift on Gold Hill.

PIKES PEAK CREEK

Perhaps as much as half of the gold produced in the Pioneer district, before the beginning of dredge mining in 1934, came from the gravel in terraces, called "bars" by the miners, on the sides of Pikes Peak Creek (pl. 11). In a single season gold worth $140,000 is said to have been recovered from one of the terraces known as Batterton Bar.

As indicated by exposures around the mine pits, the gravel in most of the terraces is not more than 5 or 6 feet thick and in Woods Flat was hardly thick enough to conceal the bedrock formed of a clay in in the Tertiary "lake beds."

In Batterton Bar the gravel consists largely of angular to sub-angular cobbles and boulders of quartzite and flintlike metamorphic rocks (hornstones). A few rounded cobbles of granite are included. Most of the cobbles are 8 inches or less in diameter. The boulders are moderately abundant. A few are as much as 4 feet in diameter but most are less than 3 feet. In the lower terraces the gravel successively becomes somewhat finer textured and more water-worn. Granite is more plentiful and scattered boulders of diorite are present. The gravel is well washed and sorted throughout. The gold from Batterton Bar occurred mostly as small particles. A few nuggets were found worth as much as $10.00 each. In trade the gold was valued at about $17.75 per ounce. Gold from Woods Flat purchased by the Larrabee Bank was 900 fine and therefore worth $18.60 per ounce.

Below the area of glacial moraines, a narrow deposit of stream gravel extends along the channel of Pikes Peak Creek. Part of the gravel on bedrock forms a pay streak that was mined in places by drifting. As time went on the partly mined alluvium became buried under tailings from mines in the terraces. Later a stretch about 3,000 feet long of the partly mined gulch gravel with its overburden of tailings was reworked with a dragline shovel and washer, in a stretch about 3,000 feet long below Treadwater Bar (pl. 11). As exposed by this working the material mined is largely composed of cobbles 8 to 10 inches in diameter, and boulders 1 to 2 feet in diameter to a depth of 8 or 10 feet.

An unknown but considerable amount of gold is reported to have been recovered during the drift mining. The later operation is said to have been profitable.

On the east side of Pikes Peak Creek an old mine extends from the above mentioned working on an alluvial cone to the top of the slope and slightly into the deposit of early drift on the east shoulder of Windy Hill (fig. 30). The drift is partly covered by the talus-covered sides of the old mine, but where exposed it is composed mostly of quartzite fragments like that of the Squaw Gulch pit on Gold Hill.

A small deposit of gold-bearing alluvium extends from the patch of terrace gravel below the early drift on Hill 5400 West, down Dry Gulch, (pl. 11) a tributary of Pikes Peak Creek. Most of this deposit has been mined but its yield in gold is not known.

PIONEER GULCH-SQUAW GULCH BASIN

A group of connected workings on K & K Bar along the east side of upper Pioneer Gulch, occupies a stretch of varying width about a mile and a half long (pl. 11). Workings in the lower third of this stretch are known collectively as the Kohrs and Kelley mine, those in the upper two thirds as the Kohrs and Bielenberg mine. Parts of this stretch beginning at the lower end of the terrace were mined in succession upstream during a period ending in 1918. The width of the area mined was from 200 to 1,000 feet and was greatest near the upper end (pl. 11). The material mined ranged in thickness from a few feet near the edge of the terrace to 50 feet toward the middle. As mining progressed upstream the excavation downstream became largely filled with tailings.

Operations in 1916 extended the workings upstream over a width of 40 to 90 feet for about 600 feet. This area is known as the 1916 pit (pl. 11). Its upstream (southeast) face, 15 to 30 feet high, exposed a bottom layer of stream gravel 2 feet in average thickness; a middle layer, 10 feet thick, of the intermediate glacial outwash that included streaks and lenses of stream gravel; and a top layer of soil, sand, and gravel. Pebbles in the stream gravel were rounded. The outwash was iron-stained and its granite boulders were softened by decay. Both the lower gravel layer and the mixture of outwash and gravel were gold bearing, but the gravel at the top was of low grade. At the southwest end of the pit near the edge of the terrace the gravel and outwash formed a pay streak 2 to 8 feet thick. The northeast face of the 1916 pit (fig. 32) exposed the rusty outwash with decayed boulders and an overlying layer of the late outwash with fresh boulders. The two layers of outwash are separated by a thin layer of sandy gravel containing a little gold. At this exposure the bottom layer of gold-bearing gravel is absent and both layers of outwash and the surface gravel are comparatively poor in gold. At all exposures either the lower gravel layer, or the intermediate outwash with or without gravel, rests on a surface eroded across Tertiary "lake beds" that dip about 15° SW. and contain scattered lenses of fine gravel with subangular pebbles. These gravel lenses contain no gold.

The 1916 pit had an area of 4,376 square yards, from which a volume of about 30,000 cubic yards was mined with a total yield of 711 ounces of gold about 887 fine. The total value of the production was $13,033, or about 43 cents per yard mined. At the present price of gold per 1,000-fine ounce ($35.00), the production value would have been a little more than $22,000, or about 73 cents per cubic yard mined. During the seasons of 1917 and early 1918 the yield per cubic yard remained about the same as for 1916, but during the latter part of 1918 appreciably leaner gravel was found, and mining was suspended.

The gold deposits along the middle and east forks of Pioneer Gulch, formerly mined by Chinese, are in stream gravels which have been concentrated from intermediate glacial drift. The gravel fills interstitial space between large boulders in the stream channels. A similar deposit, mined by Ramer and Boucher in 1935, occurs along the west fork of Pioneer Gulch. The deposit along the middle fork, mined by Conrad Warren and Andrew Moore in 1935, lies under late glacial material near the head of the stream.

Gold-bearing gravel along Pioneer Gulch, from its forks down to Pioneer, has been mined except along the west side of the channel, where a portion above French Gulch lies under a heavy overburden of talus. The late glacial outwash on the ridge between Pioneer Gulch and Squaw Gulch, is reported to contain promising material at the surface. A deposit of gold-bearing gravel, concentrated from this outwash along Schermerhorn Gulch, has been mined in one place (pl. 11).

Squaw Gulch contains a fairly large deposit composed of "lake beds," terrace gravels, and early drift that were carried down into the gulch by landslides from the slope on the north (pl. 11) and partly reconcentrated by the stream. In 1931-32 this deposit was prospected near the mouth of the gulch by the Henderson Mining Co. but no material of commercial grade was reported found.

PIONEER BAR

More than a million dollars' worth of gold (calculated at $20.67 per fine ounce) was recovered from gravel terraces mined on the spur west of Pioneer Gulch. The spur was known as Pioneer Bar. The records at the Larrabee Bank in Deer Lodge show that the gold recovered from these terraces between the years of 1875 and 1890 was from 894 to 897 fine, and was worth about $18.50 per ounce as mined.

The gold-bearing gravel occupied stream channels beneath a series of six small terraces (fig. 33) distributed from the creek to the top of the spur. A relatively small part of the total gold recovered came from material that creep had carried from the highest terrace down the west side of the spur, and into Reservoir Gulch. The transverse valley of French Gulch cuts off the terraces opposite Pioneer. Downstream the terraces end, one after another, as the crest of the spur descends to the Gold Creek valley floor.

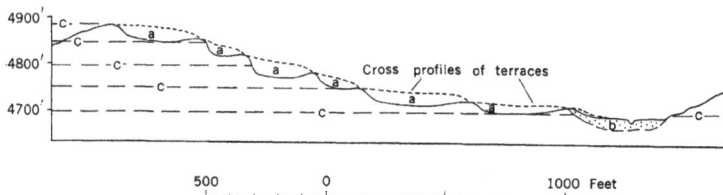

FIGURE 33.—Cross section of Pioneer Bar. *a*, Stream channel from which placer gravels were mined; *b*, Quaternary alluvium in channel of creek in Pioneer Gulch; *c*, horizontal Tertiary "lake beds."

The six terrace channels exposed by mining are from 100 to 300 feet wide and about equally spaced through a vertical height of some 200 feet (fig. 33). The bedrock is an indurated clay (Tertiary "lake beds") and the overlying terrace gravels are from 8 to 12 feet thick. The gravels contained a large percentage of subangular to well rounded cobbles less than 1 foot in diameter, and a few larger boulders. One boulder remaining in the highest channel is 3 by 4 by 5 feet in size. The rocks comprising the gravels are granite, quartzite, diorite, schist and hornstone, named in the decreasing order of their abundance.

LOWER PIONEER GULCH AND THE GOLD CREEK VALLEY

Quaternary alluvium forms a gold-bearing deposit along Pioneer Creek from Pioneer to the mouth of Pioneer Gulch and on down the southeast side of the Gold Creek valley to the Clark Fork. A part of this deposit, extending from Pioneer downstream for about a mile and a half, was mined by dredging before 1942. The dredged area ranges in width from 250 to 1,500 feet. Including an overburden of tailings, from earlier mining operations on Pioneer Bar, the deposit varied in depth from about 15 feet in the vicinity of Pioneer to about 50 feet at the lower end. Recovery by the dredge was about 18 cents per cubic yard of material mined, and the total production for the dredging operation before 1942 was about $1,200,000.

The Quaternary alluvium contains subangular to rounded cobbles, mostly 6 inches or less in diameter, that consist of granite, quartzite, diorite and metamorphic rocks. Scattered boulders are present; most of them are from 1 to 2 feet in diameter, but locally boulders as much as 3 feet in diameter are found. The alluvium was reconcentrated partly from gravel in a low terrace on the east side of the valley (pl. 11) which was undercut in places by Pioneer Gulch, and partly from the adjoining beds of late glacial outwash on the west. Near this outwash the granite boulders and cobbles are very abundant. The alluvium overlies clay beds in the Tertiary "lake beds" that form an ideal bedrock for dredge mining.

BALLARD HILL, FRENCH GULCH, AND JOBS POINT

More than $100,000 worth of gold was washed from gravels in the Ballard mine on the north slope of Ballard Hill and a smaller though considerable amount recovered from deposits in French Gulch and Jobs Point nearby.

The Ballard mine gravel is the remnant of a terrace deposit concentrated from the early glacial drift, a remnant of which underlies an area on the slope above. Both the gravel and drift contain many boulders and cobbles of diorite and andalusite schist, a fact which shows that the source of the drift and also of the gold is the basin of upper Gold Creek. Glacial striae (fig. 34) are preserved on many of the boulders. The gravel is mostly from 6 to 10 feet thick in the Ballard mine and much thinner on Jobs Point and in French Gulch. It is largely composed of subangular cobbles and boulders. Many of the cobbles are from 6 to 10 inches in diameter and most of the boulders 2 feet or more.

RESERVOIR HILL

A deposit of terrace gravel on Reservoir Hill is of similar composition to the Ballard Hill gravels but it is thin and patchy and its

FIGURE 34.—Glacial striae on a chip from a boulder of andalusite schist in the early glacial drift on Ballard Hill.

cobbles and boulders average somewhat smaller. The production from Reservoir Hill is unknown, but is probably not large.

MILL CREEK

In the valley of Mill Creek gold-bearing gravel, 20 to 30 feet wide and 3 feet in maximum thickness, is exposed in the Orphan Boy mine (pl. 11), resting on a clay (Tertiary "lake bed") bedrock. The gravel is overlain by intermediate drift composed of brownish clay and sand containing cobbles and large angular fragments of quartzite. Gold washed from this mine is rough or slightly waterworn, its particles ranging in size from pinhead to wheat grain. Nearby a similar deposit with large subangular boulders of quartzite and an irregular pay streak had been mined from a cut about 250 feet long. Old workings indicate that the gold-bearing gravel-drift deposit extends for a considerable distance below the Orphan Boy mine.

INDEPENDENCE CREEK

A mine at the head of Independence Creek penetrates the early glacial drift on the east shoulder of Windy Hill (fig. 30). The early drift is poorly exposed in the crumbled pit face and is like the drift on Gold Hill, where it is composed chiefly of an unsorted mass of angular or subangular cobbles and boulders of quartzite. The gold recovered at Windy Hill is said to have been worth nearly $20.00 an ounce (about 968 fine), and to have differed from the gold produced elsewhere in the district because it was exceptionally rough or little waterworn. This fact suggests that the gold-bearing material was plucked by the early glacier from the weathered outcrop of a gold-bearing lode in the mountains and transported to the site by the glacier. The drift has not been reworked by post-Pleistocene stream action. Below the head of the gulch alluvium concentrated from the drift has been mined. To the northeast beyond the drift area, a terrace on the northeast shoulder of Windy Hill (pl. 11) carries a thin, patchy deposit of gold-bearing gravel. To the northwest of this deposit, gold-bearing gravels have been mined from lower terraces along a branch of Independence Creek, and also from the channel of the branch.

AREA EAST OF LOWER GOLD CREEK

The area east of lower Gold Creek valley and north of Windy Hill (pl. 11) includes scattered patches of gravel, the remnants of successively lower terrace deposits. The lowest terrace, known as China Bar, was mined at one place near the former site of American Fork (pl. 11). As exposed in this working the gravel is about 8 feet thick; its cobbles and boulders are of medium or small size and are formed mostly of quartzite, granite, and metamorphic rocks and it is similar in composition to the terrace gravels in the lower part of Pikes Peak Creek, but is of somewhat finer texture. Elsewhere in this area the gravel patches are generally less than 4 or 5 feet thick and thin out toward their margins. They are medium to fine textured, and the boulders and cobbles round; the boulders few in number and mostly less than 2 feet in diameter. An exceptionally large boulder, between 2 and 3 feet in diameter, was observed on the hill near the southeast corner of section 23 (pl. 11). All these remnants of terrace gravels are either of known or probable value.

DEPOSITS IN THE MOUNTAINOUS AREA

Within the mountains, stream gravel, probably to be correlated with the terrace gravel on Jobs Point, occurs in places in the basin of upper Gold Creek. The gravel rests on a granite bedrock. In places it is overlain by, or mixed with, intermediate glacial drift. In other places it may be overlain by late glacial drift or outwash.

The gravel and the overlying glacial material from a bouldery mass (fig. 35) that has been mined by various methods in the basin area of upper Gold Creek. The principal workings from north to south are known as the Wisner, Tibbets, McFarland, Pineau, and Hughes placers. When visited in August 1949, the Master Mining Co. was operating north of the Pineau placer on McFarland, Tibbets, and Wisner property, using a dragline and bulldozer. The gravel was being washed in ground sluices with water derived from the nearby

FIGURE 35.—Glacial drift in the Pineau mine. The tunnel explores a streak of gold-bearing gravel.

lakes. The Master Mining Co.'s operation was profitable before August 1949, and many large nuggets were recovered during the summer. The average size of the gold was at least twice that of wheat grains, and many of the nuggets recovered were from 3 to 7 ounces in weight.

An irregular pit about 70,000 square yards in area and an average of about 12 yards in depth had been excavated at the Pineau placer by 1935. The pit walls expose an upper layer of late glacial drift and a lower layer of intermediate glacial drift. A layer of rusty gold-bearing gravel from 1 to 12 feet thick occurs in a discontinuous layer beneath the intermediate glacial drift. The gold from the Pineau placer was, perhaps, the coarsest mined anywhere in the Gold Creek district; one nugget from the mine weighed 27 ounces. Exceptionally large nuggets came also from all of the upper Gold Creek deposits; one from the McFarland placer weighed 24 ounces, and was on display at the Larrabee Bank in Deer Lodge where it was seen by the writer in 1916.

FORMATION OF THE PLACER DEPOSITS

The placer deposits of the Pioneer district were formed by the alternating work of streams and glaciers repeated several times since the beginning of the Quaternary period. After the uplift of the Flint Creek Range the ancestral Gold Creek and Pikes Peak Creek uncovered gold-bearing lodes as they excavated their valleys. Gold from the lodes became concentrated nearby in gravels along the stream channels. In the early part of the Pleistocene epoch, glaciers scoured the gold-bearing gravels and much waste rock from the mountain valleys, transported these materials to the benchlands, and deposited them as the early drift. In the following interglacial interval the streams deepened their mountain valleys and again concentrated gold-bearing gravels in their channels. This time, however, the gravels were not as rich as before because the lodes had become poorer in depth. In the succeeding (intermediate) glacial invasion ice again scoured the valleys, descended to the benchlands and deposited drift that was comparatively poor in gold. After a second and relatively short interglacial interval, glaciers of the third invasion severely scoured, deepened, and enlarged the headwater basins of Gold Creek and Pikes Peak Creek. The material thus eroded from the basins consisted chiefly of fresh (unweathered) granite but contained very little gold. Except for relatively small amounts left within the basins this material was transported by the glaciers to the benchlands and deposited as the late drift.

As the different drifts were deposited in the benchlands, parts of them were transported by the melt water and deposited as outwash. During the two interglacial periods the streams, as they dissected the benchlands (see pp. 85–86), concentrated parts of the early and intermediate drifts, forming the gold-bearing gravels of the different terraces (figs. 2, 3). Part of the gold in the lowest terraces came from the intermediate drift. The remainder of the gold in the lowest terraces and all the gold in the higher terraces was derived from the early drift.

In general the richest gravels (Ballard mine, Batterton Bar, p. 91) occur near areas of the early drift. A noteworthy exception is the upper part of Pioneer Bar (p. 90) which, as indicated by the rocks in its gravel, is in or near an area formerly covered by the early drift. To the north (downstream) away from the deposits mentioned, the gravels, as a rule, contain less and less gold as the distance is increased.

As the terraces were being formed, small bodies of gold-bearing gravel were left here and there in nearby gulches (Dry Gulch, pl. 11) and ravines. Existing remnants of the early and intermediate drifts enclose pockets and layers of gold-bearing stream gravel such as the blood-red gravel in the Kelley and Irvine pits on Gold Hill (pp. 86–87), pay streaks in the Orphan Boy mine on Mill Creek (p. 92), and some of the pay streaks in the K & K Bar (pp. 88–90). Apparently these bodies are parts of the gravels concentrated by streams of melt water during brief halts or retreats of the advancing glaciers. When the forward movements of the glaciers were resumed, these stream-concentrated gravels were overridden by the ice which eventually left them covered with unsorted drift.

In the Recent epoch the streams reworked parts of the terrace gravels and some of the late glacial outwash into placer deposits of medium and low grade.

ORIGIN OF GOLD

The remnant gold-quartz lodes that are exposed at the headwaters of Gold Creek and Pikes Peak Creek are the roots of veins which, before the extreme erosion of the three glacial stages, continued up into the sedimentary and metamorphic rocks overlying and adjacent to the granitic and dioritic intrusive rocks. The exposed portions of the lodes contain small amounts of gold, but it is evident that the once-rich portions have been eroded. Corroborative evidence of the diminishing amount of gold in the veins as the depth of erosion increased is the fact that the late glacial drift contains relatively little gold in comparison with the two previous glacial drifts. Only where the late glacial material has been reworked by streams is there any commercial concentration of gold.

PRODUCTION

Popular estimates of the production of placer gold for the Pioneer district run as high as $20,000,000. Dependable records are incomplete and, for the period before 1897, almost lacking. In the report of the Director of the United States Mint for 1897 the total for the district was estimated to be $4,000,000. The total production for the district to the end of 1942 is $5,667,248 distributed as follows:

Gold production of the Pioneer district [1]

Period	Ounces (1,000 fine)	Value
1897–1933 _____	12, 594	[2] $260, 318
1934–40 _____	39, 913	[2] 1, 396, 955
1941–42 _____	285	[3] 9, 975
Total _____	_____	1, 667, 248
Before 1897 (estimated) _____	_____	4, 000, 000
Total production from estimates and records _	_____	5, 667, 248

[1] Compiled from reports of the Director of the United States Mint, annual volumes of Mineral Resources published by the U. S. Geol. Survey from 1906-1924 and by the U. S. Bur. Mines from 1925-31, and annual volumes of Minerals Yearbook published since 1931 by the U. S. Bur. Mines.
[2] Calculated at $20.67 per fine ounce.
[3] Calculated at $35.00 per fine ounce.

Most of this gold came from the benchlands, the remainder from deposits on upper Gold Creek, less a relatively small amount from lodes in the headwater basins of Gold Creek and Pikes Peak Creek (fig. 26).

A total of 4,592 ounces of silver was alloyed with the 39,913 ounces of fine gold reported as recovered in the period 1934–40, mostly by dredging below Pioneer. The gold alloy mined during this period averaged about 810 fine. The records of the Larrabee Bank in Deer Lodge show that gold purchased from Pioneer Bar at times between 1875 and 1890 was from 894 to 897 fine, or, at $20.67 per fine ounce, was worth about $18.50 per ounce as mined. Elsewhere in the district the fineness ranged between 800 and 900.

RESERVES

The known placer gold reserves are limited to deposits of Quaternary alluvium along the present stream channels. The largest of these reserves is a continuation of the deposit mined along Pioneer Gulch by dredging (fig. 2). It underlies an area about 1,000 feet wide and a mile long extending down to the mouth of the creek. Smaller reserves include the partly mined deposits along Pioneer Gulch above the mouth of French Gulch, at the head of the middle fork of Pioneer Gulch, along Reservoir Gulch and in some of the ravines on Gold Hill.

Deposits classified as probable gold reserves are widely distributed in the benchlands and occur also along Gold Creek within the mountains. They include the unmined parts of the early glacial drift, most if not all of the remaining terrace gravels, parts of the intermediate glacial drift, and the alluvium in the Gold Creek valley below the mouth of Pioneer Gulch.

Most of the unmined remnants of the terrace gravel are known to contain gold. The richer of these remnants are in Pikes Peak Creek. Patches of what is probably medium- or low-grade gravel are distributed over the gradually descending slope north of Windy Hill. Down slope they merge into a sheet that includes the low terrace mined on China Bar and covers the low spur between Gold Creek and the Clark Fork. The gravel is mostly less than 15 feet thick and of medium texture. Mining of the terrace remnants depends, of course, upon the presence of water for sluicing or use of other methods for recovering the gold.

Intermediate drift exposed along the forks of Pioneer Gulch is said to yield "good pan prospects." Pay streaks like that in the Orphan Boy mine are probably to be found elsewhere in the intermediate drift along Mill Creek.

Another probable gold reserve of considerable extent along Gold Creek within the mountains is composed of stream gravel and intermediate drift.

An extensive probable gold reserve is the deposit of Quaternary alluvium in the Gold Creek valley below the mouth of Pioneer Gulch. It includes some reworked terrace gravel from the slope on the east and it has been said to contain noteworthy amounts of gold in places.

The belt of late glacial outwash in the Gold Creek valley forms a possible gold reserve. The outwash has been the subject of gold mine promotions at different times but little definite information of its gold content is available. The old dredge pit (pl. 11) mined by an English company, is said to have yielded several thousand dollars in gold. The pit is just below Pioneer Bar, but was excavated chiefly in Quaternary alluvium that is part of a gold-bearing deposit extending into a gulch just east of Reservoir Hill. The pit penetrated the underlying bouldery outwash to a slight extent only. Probably the alluvium was the source of most, if not all, of the gold recovered.

The deposit of glacial debris in which Gold Tom sank a 30 foot shaft about a mile west of Pioneer (pp. 73–74), and parts of the late glacial outwash on the median ridge in the Pioneer Gulch-Squaw Gulch basin, are also possible gold reserves.

REFERENCES CITED

BROWNE, J. ROSS, 1868, Report on mineral resources of the States and Territories west of the Rocky Mountains in 1867 : [U. S. Treas. Dept.], 674 pp.

EMMONS, W. H., and CALKINS, F. C., 1913, Geology and ore deposits of the Philipsburg quadrangle, Montana : U. S. Geol. Survey Prof. Paper 78, 271 pp.

LANGFORD, N. P., 1902, Vigilante days and ways, Chicago, A. C. McClung & Co.

MULLAN, CAPT. JOHN, U. S. A., 1863, Report on the construction of a military road from Walla Walla to Fort Benton, 363 pp., Washington, U. S. Govt. Printing Office.

PARDEE, J. T., 1950, Late Cenozoic block faulting in western Montana : Geol. Soc. America Bull. v. 61, no. 4, pp. 359–406.

STUART, GRANVILLE, 1876, Memoirs of Montana pioneers, vol. 1, Montana Hist. Soc.

U. S. BUREAU OF MINES, 1926–31, Mineral resources of the United States.

INDEX

Other Publications

by

Miningbooks.com

- Placer Gold Deposits of Nevada
- Placer Gold Deposits of Utah
- Placer Gold Deposits of Arizona
- Gold Placers of California
- Browns Assaying
- Arizona Gold Placers and Placering
- Arizona Lode Gold Mines and Gold Mining
- Dredging for Gold in California
- Metallurgy
- Gold Deposits of Georgia
- Placer Examination: Principles and Practice
- Geology and Ore Deposits of the Creede District, Colorado
- Gold in Washington
- Placer Mining in Nevada
- Gold Placers and their Geologic Environment in Northwestern Park County, CO
- Placer Mining for Gold in California
- Geology and Ore Deposits of Shoshone County, Idaho
- Gold Districts of California
- Gold and Silver in Oregon
- The Porcupine Gold Placer District Alaska
- Gold Placer Deposits of the Pioneer District Montana
- Economic Geology of the Silverton Quadrangle, Colorado
- The Ore Deposits of New Mexico

- Roasting of Gold and Silver Ores, and the Extraction of their Respective Metals without Quicksilver

- Geology and Ore Deposits of the Summitville District San Juan Mountains Colorado

www.ingramcontent.com/pod-product-compliance
Lightning Source LLC
Chambersburg PA
CBHW060448210326
41520CB00015B/3886